A Constellation Album

STARS AND MYTHOLOGY OF THE NIGHT SKY

A Constellation Album

STARS AND MYTHOLOGY OF THE NIGHT SKY

P. K. CHEN

New Track Media LLC
Cambridge, Massachusetts 2007

© 2007 New Track Media LLC
Sky Publishing
90 Sherman Street
Cambridge, MA 02140-3264, USA
SkyTonight.com

All photographs © 2007 by P. K. Chen except as noted below:
Tomb ceiling courtesy Wen Wu magazine: p.15
Spring and autumn seasonal photographs © Sandra Salamony: pp. 24 and 78
Summer seasonal photograph © Scott Brown: p. 52
Argo Navis chart by Johannes Hevelius: p. 131

Library of Congress Cataloging-in-Publication Data

Chen, P. K.
  A constellation album : stars and mythology of the night sky / by P.K. Chen.
      p. cm.
  Includes bibliographical references and index.
  ISBN 978-1-931559-38-6
1. Constellations. 2. Constellations--Observers' manuals. 3. Stars. 4. Stars--Mythology. I. Title.
QB802.C49 2007
523.8--dc22
                            2007002251
Printed in China

I would like to dedicate this book to my good friends
Leif J. Robinson, David Malin, Dennis di Cicco, Roger W. Sinnott,
Peter Sydserff, Shigemi Numazawa, Y. Takatsuki, Akira Fujii,
P. Kataoka, and K. Watanabe.

For your kindness, guidance, and assistance
with my work over the years – thank you.

TABLE OF CONTENTS

SPRING

SUMMER

AUTUMN

WINTER

Foreword by David Malin

Some years ago P. K. Chen walked into my room at the Anglo-Australian Observatory, bubbling with enthusiasm as usual. His visits were always memorable for one reason or another — he usually had some lively anecdote to tell of his recent travels or some interesting photographs to show. This time P. K. had a small book, with which he was clearly pleased. It was written in Chinese, which I do not read, but it was obvious that this was an illustrated book about the constellations. I had not seen anything quite like it before, and I was delighted to hear that it was P. K.'s own work. I strongly encouraged him to produce an English-language version; I'm pleased he followed my advice.

The stars that make up the constellations are visible to anyone who turns their eyes to the night sky, even from brightly lit suburbs. Away from cities it becomes obvious why these tiny points of light have been an inspiration to humankind for thousands of generations. Countless cultures have woven their folklore and legends into the tapestry of the sky. For example, in the West about half of our modern constellations are based on outlines that were described by Babylonian astronomers more than 5,000 years ago. Chinese astronomy also has an ancient lineage and a set of unique constellation figures that reflect its cultures of ages past.

Although today's constellation outlines play no formal role in modern astronomy, they define convenient (if somewhat irregular) patches of sky. Just as it's normal to use the names of suburbs to divide major cities — such as the districts of Chelsea and Soho in London or the boroughs of Manhattan and Queens in New York City — so too is it convenient to talk about sky segments such as Ursa Major, Leo, and Orion. The use of such ancient identities link modern astronomers firmly to their precursors who mapped the skies 50 centuries ago.

Some of the mythical beings and beasts that populate Western sky lore have no earthly counterparts. No one has seen a

8

centaur — half man, half horse. And what of a sea goat or a unicorn? Similar figures, embracing different groups of stars, appear in Chinese sky legends. Few have seen the Murky Warrior or the Azure Dragon, but most would recognize the Hairy Head — the Pleiades. Each culture projects its stories onto the seemingly random scattering of stars. The stories are not so much about the stars as they are about the beliefs of the peoples who "connected the dots" to preserve their legends.

Each constellation sky shot in this book is accompanied by a beautiful transparent overlay showing the star pattern as the ancients may have imagined it. Through the medium of astronomy, P. K.'s album brings together the past and the present, and East and West — rather like P. K. himself.

9

Introduction by
Your Stargazing Peter Pan

I was 10 when I first became interested in the stars. That was more than 30 years ago, and despite having a star map and some reference books, I didn't have much of a clue as to where to look for sights in the sky. So I stumbled along until I encountered Akira Fujii's first album that described and depicted the constellations.

In 1980, when Mr. Fujii visited an observatory in my native Taiwan, I happily had him autograph my copies of his books. For a teenager of 17 it was inspiring to obtain the signature of a tutor whom I had never met. I promised myself that someday I would create something for others as he did for me.

My first constellation album, *The Stars Above Mount Jade*, was published in 1991 and my second album, *Stars Will Be Your Friends*, appeared in 1997 (both in Chinese). Yet I was not satisfied. My desire was to introduce a new way for beginners to find their way among the stars, a way free of the struggles I encountered. I wanted to produce an introductory star guide that would be very user friendly — and I think I have done so. My "trick" is to include transparencies showing traditional constellation figures that overlay my color photographs of the sky. By doing so, the two renditions can be viewed together (and so complement each other) or examined separately.

It's important to know the locations of the key bright stars that are the soul of each constellation. It's also important to know how the overall shape of a constellation meshes with its neighbors. Many beginning stargazers have a problem seeing this "big picture," because the seeming jumble of stars overwhelms them. Actually, beginners under a dark, star-filled sky often have more trouble finding the constellations than those viewing from an urban setting, where bright city lights wash out the fainter stars!

Many of the photographs in this book were shot from the top of Mauna Kea, Hawaii.

My guide is intended mainly for stargazers in Earth's Northern Hemisphere and contains many of my favorite constellations. It's divided into four main sections, one for each season. At the beginning of every section is an all-sky chart with the same stick-figure star-pattern outlines that are present on the close-up constellation overlays that follow.

When people look at the stars in a clear, dark sky, most are struck by a sense of wonder and an irresistible urge to understand how the universe works. The first step in this endless exploration is identifying the bright stars and the constellations. I hope this book will provide a fun learning experience for children as well as grownups. If it does, I will have fulfilled my dream. I wish all skywatchers on Earth a happy voyage on the starry sea.

P. K. Chen
Your Stargazing Peter Pan

— CHEN

Getting Started

If you go outside tonight and look up, you will see the stars of your ancestors. Over time we invent different constellations — the patterns in the sky — to help us find our way around the heavens, but the stars themselves seem to remain unchanged (and unmoving) century after century. Actually, the stars *do* move relative to one another, but their motions are so slow that we don't notice the change during our lifetime. For example, Barnard's Star, a rapidly moving nearby sun, requires nearly two centuries to cross a span of sky equivalent to the width of the full Moon.

In many cases the constellations created by the Romans, Greeks, Mesopotamians, and other ancient civilizations are the same patterns we see today. The Greek astronomer Claudius Ptolemy published the first known list of constellations — 48 in all — in the 2nd-century AD though some of them, particularly those along the *zodiac* (the Sun's path around the sky), were probably created at least 2,000 years earlier. And the people, animals, and objects depicted in these celestial patterns have a rich mythological history.

Stars and Constellations

Until the early part of the last century, constellation boundaries were arbitrary, and the location of stars was not described within a single unified system. The International Astronomical Union (IAU) fixed this in 1930 when they created the modern boundaries for the 88 constellations we now recognize. (These borders are more regular than the free-flowing shapes imagined by the ancients.) The constellations are known by their Latin names, such as Andromeda or Vulpecula, which are often represented by three-letter abbreviations — And or Vul.

Many of the brightest stars have proper names — Aldebaran, Betelgeuse, Vega, Antares — which come from several languages (Arabic in particular). The bright stars in every constellation are also identified with a letter from the Greek alphabet, — the brightest is called Alpha (α), the second-brightest star is Beta (β), the third-brightest Gamma (γ), and so on. Most of the time this sequence follows the decreasing brightness of a constellation's stars, but there are exceptions. The complete name of the star is its Greek letter followed by the Latin posses-

13

sive form of the constellation's name — for example Alpha (α) Andromedae or Beta (β) Vulpeculae.

Recognizing the constellations is the first step toward becoming an amateur astronomer. Learning them is fun, especially when you discover the different stories associated with each star pattern. Apart from their beauty, the constellations and their bright stars serve as guides to the sky. Before our era of global-positioning satellites, many of the named stars were used for navigation. Today, constellation and star names remain useful for locating sights in the sky — like the names of towns on a map. It's much easier to remember Sirius than 064509-16430, the star's celestial coordinates that an astronomer might punch into a computer-controlled telescope.

Stars are classified by their brightness, which astronomers refer to as *magnitude*. The magnitude scale is upside-down, with brighter stars having smaller numbers. (The Greek astronomer Hipparchus created this odd system more than 2,000 years ago.) Thus a 1st-magnitude star is brighter than one of 2nd magnitude. The brightest night-time star, Sirius, is magnitude –1.4, and the faintest stars most people can see in a very dark sky are between magnitude 6 and 6.5.

Measuring the Sky

On Earth we use latitude (north or south of the equator) and longitude (east or west of the prime meridian located at Greenwich, England) to locate our position. Both are measured in degrees, minutes, and seconds — one degree (°) equals 60 minutes (′) and one minute equals 60 seconds (″).

Astronomers find objects in the sky using a similar system. The celestial equivalent of latitude is *declination (Dec.)*, which is measured in degrees (°), arcminutes (′) and arcseconds (″) and runs from 90° north (+90°) to 90° south (–90°).

The east-west celestial equivalent of longitude is called *right ascension (RA)*. The RA of a sky sight is determined by measuring how many hours (^{h}), minutes (^{m}), and seconds (^{s}) of time it is east of where the Sun crosses the celestial equator heading north. (This is the location of the March equinox and is currently in Pisces, the Fishes.) Right ascension is measured eastward from 0^h at the crossing point and around the sky to $23^h 59^m 59^s$.

So for example, the bright star Sirius can be found at RA $06^h 45^m 09^s$ and Dec. $-16°$ 43′. But here's a big secret. If you're exploring the sky with your eyes or binoculars, you don't need RA and Dec! Only when

14

This Chinese star map from the early 12th-century AD shows the 28 lunar houses surrounded by the 12 signs of the (Western) zodiac. This map is painted on the curved ceiling of a tomb (near Beijing) of an official who died in AD 1116. This photograph was first published in *Wen Wu* magazine.

Zodiacal Constellations

"What's your sign?" astrologers ask. The signs to which these pseudoscientists refer are 12 constellations: Aries, Taurus, Gemini, Cancer, Leo, Virgo, Libra, Scorpius, Sagittarius, Capricornus, Aquarius, and Pisces. Collectively they form the *zodiac* — the constellations though which the Sun, Moon, and planets pass during the course of a year. (In fact, these solar-system bodies also spend time in a 13th constellation — Ophiuchus — something astrologers conveniently ignore.) The *ecliptic* is the Sun's exact path through the stars of the zodiac, though it actually marks the plane of the Earth's orbit around the Sun.

Chinese Lunar Houses

The ancient Chinese constellations are much smaller than those of the Greeks — in fact many consist of only two or three stars. And unlike the Greek patterns that depict myths, Chinese star patterns were based on Chinese court life and day-to-day activities. Astronomers in ancient China noted that as the Moon moved through the sky, it passed through 28 different star patterns before returning to the same spot in the sky about one month later. So it was said that each night the Moon dwelt in a different constellation — hence the term *lunar houses* or *mansions* for these 28 constellations.

you use a telescope to search for faint objects do these celestial coordinates become important.

Instead, there's an easier way to describe angular dimensions in the sky — use your hand! Your little finger held at arm's length covers about 1° of sky, your fist about 10°, and your fully open hand — from the tip of your thumb to the tip of your little finger — about 20°. The full Moon is 30 arcminutes (1/2°) across. The maximum resolution of your unaided eye — that is, the smallest detail that it can see — is about 1 arcminute.

Celestial Motions

During a 24-hour period the Sun, Moon, planets, and stars appear to slowly move from east to west. This is an illusion caused by Earth's rotation from west to east. As Earth turns, everything in the sky swings around the *north* or *south celestial pole,* the projection of Earth's North and South Poles into space. You can see this by taking an hour-long (or more) photograph with your camera aimed at one of the poles. The stars will appear as short arcs, not dots, and the arcs will be centered on the pole (which,

15

if you're looking north, means they'll be centered on Polaris, the Pole Star). During a long winter's night, you can watch stars near the *celestial equator* (Earth's equator projected into space) rise in the east and complete their journey across the sky into the west.

At the same time the starry sky slips about 1° westward day by day. This happens because we define a day as being 24-hours long, whereas Earth actually takes only 23 hours and 56 minutes to complete one rotation. This difference of four minutes is equivalent to 1° in the sky. From day to

There's something missing in this photo of star trails around the South Pole — a bright Pole Star! Northern Hemisphere stargazers are lucky to have Polaris, the North Star, to mark the location of the North Celestial Pole. There's no similarly bright star over the South Pole.

16

Cataloging Celestial Sights

When a *deep-sky object* — a star cluster, nebula (an interstellar cloud of gas and/or dust), or a galaxy (a vast star system like our Milky Way) — is identified by an "M" designation, it means it was included by Charles Messier in his famous *Messier Catalog,* first published in 1771. The catalog now includes 109 interesting objects (some were added after his death), all of which are visible from temperate northern latitudes. This French comet hunter made his list to identify a celestial sight whose appearance in a small telescope might be confused with that of a comet. Another list containing many excellent targets for small telescopes is the *New General Catalogue (NGC)* published by Johann Dreyer in 1888. Today there is an abundance of excellent publications, websites, and software to aid stargazers who want to look for subtle celestial sights; a few sources are listed on page 132.

The stargazer's name for the Orion Nebula (in red) is Messier 42 (M42). The pretty purple glow above it, nicknamed the Running Man Nebula (can you see him?) is known as NGC 1977.

day, this slippage goes unnoticed, but over a month or so the change becomes obvious. The sky you see at 8 p.m. in mid-February will differ conspicuously from the one you saw at 8 p.m. in mid-January.

Where Are the Planets?
Including Earth, there are eight planets in orbit around the Sun. In order out from the Sun they are Mercury, Venus, Earth, Mars, Jupiter, Saturn, Uranus, and Neptune. (In mid-2006 the IAU decided that Pluto was not a regular planet but was, instead, a "dwarf planet.") Neptune is very difficult to see with the unaided eye, and sighting Uranus requires a dark sky and a good finder chart. All the planets change their position in the sky, some rapidly and others, especially the outer two, slowly. That's why they're not included on my constellation charts. To find out where the planets are on any night, check an astronomy yearbook or

17

magazine. A good web site with a star chart that you can customize for your location is SkyTonight.com — it will show you where the bright planets are to be found.

Meteors and Meteor Showers

Meteors are streaks of light typically lasting a fraction of a second. They occur when particles the size of tiny pebbles or sand grains are incinerated as they plunge through Earth's atmosphere. When these particles are drifting through space they're known as *meteoroids*. If a meteoroid survives its fiery plunge through the air and lands on the ground, it's called a *meteorite*. When many meteors fall from the sky during a short period of time, it's called a *meteor shower*.

Most of these particles are fragments shed by a comet as it orbits the Sun. Don't confuse the two: comets are slow-moving and may be seen for weeks or months with the unaided eye or a small telescope, while meteors are visible for mere seconds as they pass through our atmosphere.

Many people believe that meteors, or shooting stars, are a sign of good luck and an opportunity to wish for something. So looking for a meteor, or watching a meteor shower, is fun and may be good luck, too!

Choosing the right place to watch a

meteor shower is very important. It is wise to find the darkest place you can, for even a little extra light in the sky (light pollution) will seriously reduce the number of faint meteors you can see. And don't forget to

check the phase of the Moon. If more than a sliver of Moon is in the sky, the shower will be compromised.

From astronomy almanacs, magazines, and web sites you can learn when an an-

During a meteor shower, you may see a meteor a minute, or more. But you need to be well away from city lights to spot this many.

nual shower will peak and how strong it is likely to be. Not all showers are alike. Some are over in less than a day while others last for weeks; some have slow-moving meteors, others fast-moving ones. The most active showers are the *Quadrantids,* which peaks for a few hours around January 4th, the *Perseids* around August 12th, and the *Geminids* around December 13th. These names derive from the constellation that hosts the *radiant* — the point in the sky from which the meteors seem to flow. (The Quadrantids are from a now-defunct constellation; these meteors actually radiate from Boötes.)

When you go out to observe, it's important that the radiant be as high in the sky as possible because the lower the radiant, the fewer meteors will be seen. Also, you'll see more meteors after midnight than before. This is because after midnight the part of Earth you're observing from is facing forward, so you're better positioned to see debris plowing through the atmosphere than before midnight.

Also watch out for the occasional really bright meteor called a *fireball.* These are usually random meteors, though sometimes a meteor shower produces a fireball or two.

Tools for Stargazing: Binoculars

A major misconception is that you need a telescope as the next step beyond observing the sky with your unaided eye. Not so. A wonderful instrument that you may already own is binoculars. Any model will work and

The next time a meteor shower is predicted, get some friends together, find a dark-sky site, and have a meteor-watching party.

Binoculars are great for stargazing because they can be used anywhere, at any time, and there's no setup required. Take them along the next time you go camping.

19

give you a nice view of many of the sights mentioned in this book. Binoculars also let you see a much wider swath of the sky than a telescope can and are thus wonderful for examining the rich star clouds of the Milky Way and large sky sights such as the Hyades and the Pleiades Star Clusters in Taurus. The performance of binoculars, especially if they're heavy or are of high magnification, can be dramatically improved by mounting them on a steady tripod.

A 7 × 50 binocular is one that magnifies 7 times and has a 50-millimeter-diameter *aperture* (the diameter of the front lenses). It's often cited as best for astronomical observing, but that's not completely true, especially for older adults. The pupils of old eyes usually cannot open wide enough to capture all the light that comes out of the eyepieces of such a binocular. In fact, I suggest that the best stargazing binocular is a 10 × 50, though you will probably need to mount it on a tripod to keep it steady.

Tools for Stargazing: Telescopes
Of course, sooner or later you'll probably want a telescope. Today there are scores of scopes and dozens of manufacturers and distributors from which to choose. It's beyond the capacity of this sky guide to go

Telescopes give fine close-up views of celestial sights, and if you become deeply interested in the universe, you'll probably want to get one eventually. But you don't need a scope to enjoy this book.

20

into detail, but a good-quality *refractor* with a main lens that's 3 inches (75 millimeters) in diameter or a *reflector* with a primary mirror at least 6 inches (150 mm) across can be purchased for a few hundred dollars. That's the aperture range I'm thinking of when a "small telescope" is mentioned in the text.

So large is the variety of instruments available that if you're a first-time buyer, you should carefully study telescope test reports and advertisements in astronomy magazines and on web sites (see page 132) to determine which seems best for you. Better yet, join an astronomy club to get lots of advice and a chance to examine and look through a variety of telescopes firsthand before buying.

One more thing. Binoculars show the sky with the same orientation as you see with your unaided eye. But astronomical telescopes sometimes turn the sky upside down. Even east and west can be reversed, depending on the type of telescope or the accessories used. That's why I like observing with my eyes or binoculars. Besides, you can't see all of a constellation at one glance through a telescope!

Tools for Stargazing: Other Necessities

Binoculars and telescopes are not the only observing aids a stargazer needs. In addi-

Observing Necessities

In addition to a star map and a red-light flashlight, consider taking some or all of these items with you when you head out for a night of skygazing.

1. List of objects you want to see
2. Planisphere
3. Binoculars (even if you have a telescope)
4. Warm clothing and a windbreaker (even in summer)
5. Reclining chair or outdoor mattress (or a blanket)
6. Water or other (non-alcoholic) beverages
7. Nutritious snacks
8. Bug spray

When I do serious deep-sky observing with my telescope, I need large-scale charts. One of my favorites is *Sky Atlas 2000* (from Sky Publishing). Also, I don't go anywhere without my red-light flashlight.

tion to the photographs and constellation overlays in this book, you'll want a star map. There are four seasonal charts in this book (on pages 25, 53, 79, and 105). They'll get you going, but you might need more.

If you're not at all familiar with the night sky, start with a *planisphere*. This is a sky map on a moveable disk that can be set to a particular date and time. The sky above your horizon is shown through a window in the disk. To be most useful, a planisphere must be designed to match your latitude to within 5° or so. They're available as traditional paper products, software, and on the Web. An excellent, interactive one can be

21

found at SkyandTelescope.com.

Once you know the constellations, move on to a detailed sky map. The *Bright Star Atlas* by Wil Tirion (Willmann-Bell) covers the entire sky in 10 charts; it's good for wide-field views of the heavens. A physically smaller book containing 80 detailed charts is the *Pocket Sky Atlas* by Roger W. Sinnott (Sky Publishing, 2006). There is a great variety of other printed sky atlases and charts available as well as excellent software and information on Web sites; more are listed in the Resources section on page 132.

Another essential tool for stargazing is a proper light for reading a star chart or reference book. The best views of the sky happen when your eyes are *dark adapted* — that is, when they have not been exposed to bright light for at least 15 minutes. If you need light, the easiest way to preserve your night vision is to cover a small flashlight with a sheet of deep-red transparent film. Or, for a few dollars you can buy a small red-light LED flashlight. In either case, keep the light level low.

Six Steps to Great Stargazing

1. Wait for good weather. Clear skies often follow the passage of a cold front. A deep-blue daytime sky is an excellent predictor that the night will be superb. Even a partly cloudy sky can be frustrating — you'll find it difficult to orient yourself to the stars, and even if you find a celestial object, your view will be frequently interrupted.

2. Know what's up and what you'd like to see. You can't observe Orion after sunset in July; you'll have to wait until winter. That's why a planisphere is useful — it shows you what's up during the particular evening you want to go out and stargaze.

3. Be sure to check the phase of the Moon and when it will rise or set. One of the most convenient times for skywatching

Light pollution makes it difficult for urban dwellers to see the stars, so find a dark-sky observing site away from city lights. Also, try to star gaze when the Moon isn't in the sky.

22

is between new Moon and first quarter because the moonlight will not be too bright, and you can go to bed at a reasonable hour. When the Moon is full, its light washes out the fainter stars. Another good time is after last quarter when the Moon rises late.

4. If you're traveling to a dark-sky site, depart early to give yourself time to set up your observing station before evening twilight fades. I find it very relaxing to watch the stars appear as night descends; it helps me forget the troubles of a hectic day and focus on the observing task at hand.

5. Knowing the cardinal directions will help you orient yourself to the sky. If you lift your left arm straight out from your side and point it at the Sun as it sets, you will automatically be facing north (at least roughly). As the sky darkens, try to locate a bright star or star pattern (the Big Dipper, for example), then find more constellations as more stars appear.

6. You'll likely recognize a constellation soon after you've spotted one or more of its brightest stars. Use the photographs and transparent overlays, or the seasonal star maps in this book, to locate the other members of the star pattern. Soon you'll find stargazing as effortless as making a cup of tea or coffee.

Please Observe!

I hope my book of constellations encourages you to go outside many nights and look up. The stars are a magical sight and often inspire wonder and curiosity. If this album helps stir your curiosity, then I will be satisfied.

Happy stargazing!

Spring Sky: Highlights

Face north in the evening to find the Big Dipper, the most famous and recognizable of all the star patterns visible from Earth's Northern Hemisphere. For those viewing from midnorthern latitudes, its seven bright stars, looking like a big spoon, will be hanging upside down high above the horizon. The Big Dipper is part of the constellation of Ursa Major, the Great Bear, as hinted at by the stick figure on the chart opposite. (Can you see the Bear? If not, turn to page 32.)

Now, follow the arc of the Dipper's handle southward until you spot the bright yellow-orange star Arcturus. It's the alpha star in the constellation of Boötes, the Herdsman. Keep going southward (now it's best to turn around and face south). At about the same distance as Arcturus is from the tip of the Dipper's handle is another bright star, blue-white. It's the constellation Virgo's alpha star, known as Spica.

To the lower right of Spica you will find a small, rather dim though conspicuous, trapezoid. It marks Corvus, the Crow, a noisy fellow. Observers lucky enough to be south of latitude 25° north can use Corvus as a guide to the four brightest stars of Crux; they form the famous Southern Cross, which lies directly to the south (and below the horizon on the all-sky chart opposite). To the left of Crux are two very bright stars, Alpha (α) and Beta (β) Centauri, also known as Rigil Kentaurus and Hadar, respectively. Centaurus, of course, is a centaur — half man and half beast.

Moving back to Arcturus, a line directed toward the west will lead you to Regulus, the alpha star in the constellation of Leo, the Lion. Together with Spica, this stellar trio forms what is sometimes called the Spring Triangle.

After evening twilight fades you can see Gemini, the Twins, heading down into the west. It's as if they are saying: "Farewell, my friends. See you next year!" The winter Milky Way runs low across the western and northern horizons, where it's dimmed by atmospheric haze; on spring nights you cannot see it well. Save your Milky Way viewing for the summer months to come.

This star chart is most accurate if used within an hour or so of the times listed and is plotted for observers located between 30° and 50° north latitude. All times are standard time; if daylight-saving time is in effect, add one hour.

To use this chart, hold it in front of you and rotate it so that the yellow label corresponding to the direction you are facing is positioned at the bottom, right-side up. The stars in the sky should match those depicted on the chart. The center of the chart is the zenith, the point in the sky directly overhead.

24

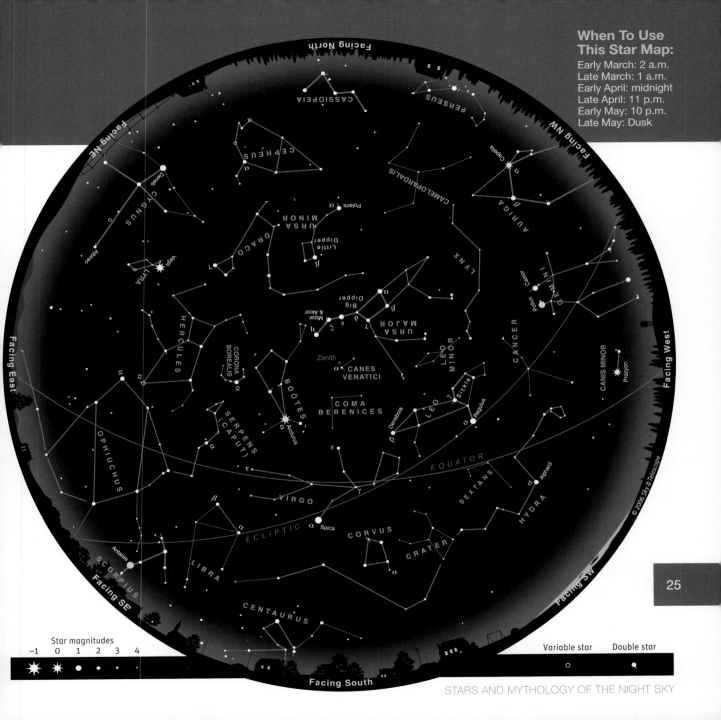

When To Use This Star Map:
Early March: 2 a.m.
Late March: 1 a.m.
Early April: midnight
Late April: 11 p.m.
Early May: 10 p.m.
Late May: Dusk

Facing North

Facing NE

Facing NW

CASSIOPEIA

PERSEUS

CEPHEUS

CYGNUS

Deneb

Albireo

CAMELOPARDALIS

Capella

AURIGA

α

β

LYNX

Polaris α

URSA MINOR

Little Dipper

β

DRACO

Vega

LYRA

GEMINI

Pollux   Castor

Big Dipper

Mizar & Alcor

α

δ

ε

ζ

η

γ

β

URSA MAJOR

LEO MINOR

CANCER

CANIS MINOR

Procyon

Facing West

Facing East

HERCULES

CORONA BOREALIS

α

Zenith

CANES VENATICI

α

LEO

Sickle

γ

COMA BERENICES

Denebola β

α Regulus

OPHIUCHUS

BOÖTES

SERPENS (CAPUT)

Arcturus α

3

EQUATOR

SEXTANS

α Alphard

HYDRA

VIRGO

τ

β

γ

ECLIPTIC

α

α Spica

CORVUS

CRATER

α

Antares

SCORPIUS

LIBRA

Facing SE

CENTAURUS

Facing SW

25

Facing South

Star magnitudes
−1 0 1 2 3 4

Variable star   Double star

© 2006 Sky & Telescope

STARS AND MYTHOLOGY OF THE NIGHT SKY

Stories About the
Spring Constellations

In March the Sun, heading northward in the sky, crosses the celestial equator (the projection of Earth's equator into space). When it does so — on the day called the *equinox* — day and night are approximately equal everywhere on our planet. In the North Temperate Zone memories of the short days of winter are beginning to fade, and the temperature is slowly rising. Our land is returning to full life during this exciting season.

On April evenings you can still see a few of winter's constellations and bright stars hanging around in the west. But their party is over! A new spring constellation has moved onto center stage. Ursa Major, the Great Bear, now prances nearly overhead. But not all constellations glitter equally. The Great Bear's junior partner, Ursa Minor, the Little Bear, hangs well above the northern horizon, while east of Ursa Major, Coma Berenices will challenge you to find it!

In May, hints of summer are coming to northern latitudes. In gardens, the earliest spring flowers have already faded, but high in the mountains the blossoms are just beginning to appear. Swallows play hide and seek amid puffy clouds. The Sun has slowed its rapid northward climb. Temperatures are rapidly rising; daylight hours are nearly at their peak and nighttime hours at their minimum.

Yet, on a May evening, you can still briefly see some winter constellations such as Auriga and Gemini in the west, while spring constellations such as Boötes and Virgo hang near the *meridian*, the imaginary line that runs through the heavens from north to south (through the *zenith*, the point in the sky that's directly overhead) and divides the sky in half. Draco has risen higher in the northeast, yawning and stretching as it gets ready to follow in the steps of Hercules.

Lower in the northeast two birds take wing — Cygnus, the Swan, soon to be followed by Aquila, the Eagle. Also there gleams Lyra's Vega. Far to the south, Centaurus waves and says, "I'm here, don't forget me!" And for some, this is the season to seek the famous Crux (Southern Cross), which lies below the Centaur.

Lying just east of Boötes, the Herdsman (see page 42), is a pretty half-circle of stars — Corona Borealis, the Northern Crown. It's said to be the crown of princess Ariadne of Crete, tossed into the heavens to commemorate her wedding.

Corona Borealis

BOÖTES

Arcturus

Leo

Leo, the Lion, sports one of the few constellation patterns that truly resemble its namesake. The distinctive **Sickle** (or mirror-reversed question-mark), punctuated by 1st-magnitude **Regulus — Alpha (α) Leonis** — neatly mimics the head and forequarters of a crouching feline. Also, Leo's trailing triangle, with its 2nd-magnitude beta star **Denebola**, resembles the haunches of a cat about to spring. Incidentally, ancient Chinese farmers regarded Leo as the Rain Dragon, and the appearance of the Lion's Sickle in the eastern evening sky was an indication that spring rains were imminent and it was time to sow seeds.

Regulus, which means "Little King," is a fitting appellation for brightest star in the celestial King of Beasts. The star lies near the ecliptic, 78 light-years from Earth. (A light-year is equivalent to 5.8 trillion miles or 9.5 trillion kilometers.) At magnitude 1.4, Regulus is the 21st brightest star in the night sky.

Just north of Regulus lies the bright star **Gamma (γ) Leonis**, which holds a surprise for observers with a small telescope. Also known as **Algieba**, which means "Lion's Mane," Gamma is actually a pair of stars that appears as one to our unaided eyes. But with a telescope the star can be split into its two yellow components of magnitudes 2.4 and 3.6. These stars are separated by 4½″ (arcseconds), a tiny amount equivalent to 1/400th the diameter of the full Moon. The fainter star lies to the southeast of the brighter one. Gamma is one of the finest double stars in the entire sky. Furthermore, it's a binary; the two components circle each other every 620 years or so.

Gamma also marks the radiant where in mid-November, at internals of roughly 33 years, the famous **Leonid meteors** emerge to provide a spectacular display. At their best as many as 200,000 meteors may streak across the sky in an hour. You can still see a modest meteor shower every year while waiting for the next Leonid storm around 2032.

In mythology, Hercules was commanded to kill the Nemean lion, represented by the constellation Leo. (This was the first of Hercules' "Twelve Labors.") He succeeded. Ancient Persians believed Regulus was one of four Royal Stars guarding the sky (the others were Aldebaran in Taurus, Fomalhaut in Piscis Austrinus, and Antares in Scorpius).